Song of Two Worlds

Song of Two Worlds

Alan Lightman

CRC Press
Taylor & Francis Group
Boca Raton London New York

CRC Press is an imprint of the
Taylor & Francis Group, an **informa** business

AN A K PETERS BOOK

CRC Press
Taylor & Francis Group
6000 Broken Sound Parkway NW, Suite 300
Boca Raton, FL 33487-2742

First issued in paperback 2017

CRC Press is an imprint of Taylor & Francis Group, an Informa business

No claim to original U.S. Government works

ISBN 13: 978-1-138-11676-4 (pbk)
ISBN 13: 978-1-56881-463-6 (hbk)

This book contains information obtained from authentic and highly regarded sources. Reason-able efforts have been made to publish reliable data and information, but the author and publisher cannot assume responsibility for the validity of all materials or the consequences of their use. The authors and publishers have attempted to trace the copyright holders of all material reproduced in this publication and apologize to copyright holders if permission to publish in this form has not been obtained. If any copyright material has not been acknowledged please write and let us know so we may rectify in any future reprint.

Visit the Taylor & Francis Web site at
http://www.taylorandfrancis.com

and the CRC Press Web site at
http://www.crcpress.com

Library of Congress Cataloging-in-Publication Data

Lightman, Alan P., 1948–
   Song of two worlds / Alan Lightman.
      p. cm.
   ISBN 978-1-56881-463-6 (alk. paper)
   I. Title.

PS3562.I45397S66 2009
811'.54–dc22

                                                          2009007630

Cover photograph courtesy of Daren K. Jaeger Photography.

Photograph on page 38 copyright © Christina Craft.

Thou hast made me endless, such is thy pleasure.
This frail vessel thou emptiest again and again,
and fillest it ever with fresh life.

—Rabrindranath Tagore, *Gitanjali*

Part I

Questions with Answers

1

Awake –
What are these quick shots of warmth,
Fractals of forests
That wind through my limbs?
Fragrance of olive and salt taste of skin,
Razz-tazz and clackety sound?
Figures and shapes slowly wheel past my view,
Villas and deserts, distorted faces,
Children, my children –

2

Distant, the pink moons of my feet.
What rules do they follow?
I think movement, they wondrously move,
Moons flutter and shake.
I probe the hills and the ruts of my face –
Now I grow large, now
I grow small, as the waves
Of sensation break over my shore.

There, a gnarled tree I remember,
A stone vessel, the curve of a hill.
What is the hour?
Some silence still sleeps
In my small sleeping room –
Is it end or beginning?

2

Have I awakened?

For decades, it seems, I have slept in a cave,
Hung like a dried fly
That's sucked of all faith.
Am I awake
After so many foldings unfoldings,
Grown numb to the world of belief?

Something is stirring, some newness,
A flail, buzz, and heave.
Welcome, this sharp morning blast –

Pleasure floods through me
While tears sting my eyes,
Veins fill with promised life.
Breathing, I breathe and I feel,
My skin bristles.

3

Footsteps –
It's Abbas, dear Abbas.
I know that old shuffle,
Grey stubble, haired mole,
And the yellowing teeth.
Clatter of pots in the kitchen.
He's making some tea.
"Are you awake?" he roars.

Smells of hot peppers and onions
With cinnamon, hazelnut,
Baklava, sugared cream.

I rise from my bed, middle aged,
Balding, the white scar on my arm,
Shrunken chest,
Losing more weight every year –
In thirteen, by my estimate, I'll weigh zero.
My spindly legs stiff as I stand,
Light from the night hallway,
Red glint of my eyes.

*   *   *

4

Am I still sleeping?
I dreamed of my uncle Zafir,
Weighing the sand on the beach.

4

Abbas is muttering.

Standing, I look for my paper and pen,
Books lying scattered about. Inhale –
I breathe in my ancestral home,
Turquoise rough stucco and terra cotta tiled floors,
Earth colors, arches and airy rooms,
All crumbling now. There, the tinny piano
My mother once played. Here, the brass compass.

Abbas serves breakfast,
Eats at his small bench,
Belching and smiling.

Through an arched window,
I gaze at the wide rutted steps
To the terrace and down to the sea.
Garden of aloe and sharpened spine puyas,
The dune evening primrose, the prickly white poppies,
The red bougainvilleas that wind up the walls –
Shadowy shapes in the dim light of dawn.

   *   *   *

There, bitter orange trees,
Now smelling vanilla and powdery.

Olive groves, gift of my father,
Like everything here.

Parentless now. I was a parent myself,
Father and husband.

5

Then faintly, the call of the muezzim,
The nasalized song.
Abbas drops to the floor, praying.

I watch him and wait,
Help him regain his feet,
Give him his cane.
I am blasphemy.

"Shukr," he says, rubbing his bony knees.
Glances at me, sighs,
Hobbles from the room.

6

I take up my pen, dry for some years.

What should I write?
What should I think?

6

Escape from this slow daily drip –
Keeping accounts, trips to the tailor,
My sweeping cracked plaster, the buckling
Unbuckling of sandals,
The sippings of tea,
Life without life –
Drivel from dry sea beds,
While time slides to an end.

But something has turned, opened,
Some wrinkling of air, brain cell that shuddered,
Perhaps old Zafir called from the grave –
And the tree is no longer a tree,
Hand is no longer a hand.
I won't sleep to the end.

Look, morning light on my desk –
I will take up my pen.

7

In the distance, I see a great tower,
A built thing, a place.
Child, I'm a child, song of discovery –
I'll move and explore.

There, a long hallway,
A dome, silent stars
Splitting the blackness from light –
  *   *   *

I knock on the door of the universe.
Here, this small villa, this table, this pen.
I ask the universe: What? and Why?
Now wakened, I must remake the world,
One grain at a time.

8

A great door opens a crack,
Sends a shaft through the darkness,
Which brightens one tile in the hall,
And I want to see more, and forever.
In the dim hallway, dark galaxies
Spin in the space of my mind –
Secrets of matter

8

Now wait in the dark for the opening door –
And the secret of time waits in the dark
For the opening door –
And the puzzle of death waits in the dark
For the opening . . .

What power pries open the door?
Is it divine spirit or deity?
Standing, I wait for some thing to believe.

9

Past the half open door lies the sea –
Fishing boats bucking on anchors,
Gulls wading for clams, wind on my skin,
Something that gleams on the shore,
Sliver of glass. Stooping, I grasp it,
It catches the light,
Casts off a palm's width of colors,
The reds, greens, and violets, the pattern of rainbows.
A second shard does just the same, and another,
Another, always one pattern.
And this I believe.

10

Many hours I spent with Zafir
In his chemical shop

Smelling of sulfur and salts,
Burnt oil and dry ice.
Cannisters hissed with blue opaline clouds –
Curved glass and scales.

"One thing is certain, life flies,"
He said, quoting Khayyam.
"And here is another:
This manganese, density 7.2,
Silver in tint, atomic mass 54.9.
Truth – it is truth."
And he placed the cool bar in my hand.

11

As I hold this glass shard in my hand.

I have entered the cosmos of questions with answers.
This is the world of the sharp spheres of hail,
Orbits of planets, vibrations of atoms,
The fission of cells, pulse of a neuron,
The plucked string of a harp,
Wavelength of blue light.

Here, atoms are pierced
By equations, the sunlight exposed by a prism,
The cells observed passing their secrets
Of organ and bone. Nothing escapes

Being weighed and titrated, resolved
Into numbers, displayed in a graph.

Is this the place where I make something whole?
Find a solidity?
Slow down my dark fall to nothingness?
Subdue the voice that says No?

Should the lab be my temple?
The microscope my prayer mat,
The stopwatch my candle?
My sacraments test tubes and beakers,
Pipettes, calorimeters,
Specimens, diodes,
Transistors, glass flasks?
Should my guides now be Darwin, al-Haytham,
Lavoisier, Einstein, and Newton?

12

I knock on the door of the universe, asking:
What makes the light of the stars?
What makes the heat of my flesh?
What makes the tear shape of rain?

Questioned, the world walks on little feet.
Ask and take wing while my body
Stays bound far below –
Body, poor stone
That will wear down

To dust,
Like this ruined villa perched on the sea.

So much that I've lost,
I have nothing
Except a fierce hunger
To fathom this world.
Naked, I knock on the door,
Wearing only my questions.

13

Morning breaks in the east,
And my house fills with light.
I discover my various things:
Tables and chairs, closets of clothes,
Drawers full of coins, shelves of books.
Still my house seems abandoned and bare.

Somewhere, the market is seething.
Men rush down the alleys
And rush to their deaths.
Horns clamor and wail. City jerks
Screaming, my mind becomes deaf.

I walk past the temple that glows –
Here, I own nothing,
Except this unease.
And the thought of the ages lifts

This poor stone and says:
"Here lies the path to eternity."

14

It is time to discover the oceans –
I fling myself into the deepening,
Roll and slap, swallow what fits in my maw,
Launch my small skiff on the thrashing green sea,
Follow the bell buoys of my mind.

A dot in the roiling,
Convulsed by the waves,
I ask and find stillness in knowing:
This is the stuff of the water,
And this is the shape of the sky.
This is the measure of masses,
And this is the logic of air.

Adrift in this strange jittered world,
I am held by these meters and grams,
By Zafir's atom densities,
Wind on my sail,
Action reaction,
The compass submitting to magnetic force.
These are the bits of some pattern,
I think, and I hold and am held,
And the numbers are wind and rock,
Even a holiness
Holding my body, and this might be the way.

15

Abbas brings an orange from the grove,
Peels, cuts, and peppers it,
Swallows a piece and gives me another.
I flinch at the bitter and raw,
Spit out the pulp.

"I'm out for the pruning," he says,
Sweeping the peel to his pocket.
"You spend too much time alone."

"See if a letter has come," I ask,
As I ask every morning,
My futile and vain prayer for the day.

He nods his head delicately,
Touches my arm
With his veined, mottled hand.
"Orange pudding I'll make later this week."

"What day is it?"

"Tuesday."

16

Great Al-Haytham –
Believer in methods of science, the meters and grams,
Finder of Fermat's "least time" before Fermat,
Practitioner of "Galilean experiments" before Galileo,

Feigner of madness to escape
Death by caliph –
Great Al-Haytham,
Show me your faith.

You found the way that we see:
Light starts from beyond the body,
Then enters the eye,
Slanting through crystalline humor
And vitreous gels.
Light does not streak from the eye,
As others thought.

With sight tubes
And taut string and chambers,
You found out the movement of light –
Rectilinear, reflection, refraction.

Mathematical edge of a knife –
You found the point on a surface
That connects other points via light,
Like the point in your mind
Joined to the sharp point of truth,
And the one thing
A hand must hold.

There, at your table, with rulers, protractors,
Warm breeze from the gulf, goats crapping –
Al-Haytham, tell me what I can see –
Is it this sun in my room?

17

Smaller, the room I had decades ago
When I went to the country across the sea,
Taking my uncle's old chemistry books,
Slide rules and weights,
Studying science, philosophy, poetry –
Churches for mosques,
Trousers for caftans,
Wood creak of desks, thick fingered tutors
And beveled glass, laughter of students,
Rectangular courtyards of grass –
Never I felt at home.

18

In the blue twilight
I wander my groves
While the wind whispers softly,
The puyas sway sleepily, waiting for night.

I gaze at the sky
To admire a hoopoe
And pink rising moon.

In the blue twilight, I gaze
At the bird and the papery moon –
One throbbing flesh and the other cold dust –
And I marvel at what harmonies
Couple the two.

Electrons in orbits contained by the quantum,
And atoms conjoined by the dipolar force,
Gravity balanced by pressure
And flap of a wing.
Here, I remember the angles and curves,
Calculus, I learned it all –
Each bird bit and moon fracture
Fixed by the symmetries,
Airborne and airless, aloft.

19

A speck in the distance, across the burnt sand.
"She's coming," Abbas says, "the laundress's daughter."

"Then it's Wednesday."

"In five minutes she'll be at the gate."

"Ten minutes," I say, holding my thumb out
To measure her angle. "I reckon
She's one thousand meters away."

"Your mind is as tight as a sheep's ass."

Ironed caftans and shalwars,
Our white cotton cloth,
Even Abbas' ragged boxers, all clean.

    *   *   *

The laundress's daughter waits shy
On the terrace, she's eighteen or nineteen –
Abbas counts out coins in her hand
Without touching. I look from my window,
A dark strand of hair has just slipped
From her scarf, rests on her shoulder,
And somewhere the groan and the love smell,
My cock rising vertical, lifetimes ago. Quickly
She tucks back her hair, turns away,
Softly says, "Pick up on Monday."

20

Bowl of couscous. I eat
But I cannot be filled.
On my table the couscous, an inkwell, a pen.
And a photograph kept in a drawer –

A woman I barely remember,
A prickle of life left only here, in this picture,
Her days and nights vanished. What string
Of knots led to our meeting that day
On the bridge? What caused her to look
At me, then look away? Was it the slant
Of the sunlight, the hue of my hair?
Why did she speak to me?

How many nights
Did she sleep in my bed? I've forgotten
The sound of her voice,
Tilt of her head,
The calm sea of her hand.

Sometimes I dream that the night dancer
Drifts past my bed.

And the others, my children and wife.
Did I have any choice?

I want rock and hard edge, gush
Of my blood in straight lines,
Matter, inertia, and force,
Explanation.

21

Newton, thin as a mathematical line,
Aquiline nose, fierce eyed and secretive –

You kept a notebook of questions,
The dip of your quill in an ink of oak galls.

You sought the gods' hidden knowledge,
Sketched in that notebook your pictures
Of pendulums, levers –
You pondered the sloping of curves,
Tangents and lines –
Toyed with your series of numbers
That marched on forever like waves of the sea –
Thought of all motion as tiny infinities,
Fluxions and calculus.
You found equations that govern
The comets, the path of the moon, gravity,
Shape of the stream in my bath.
You split the colors of light.
"All can be known."

You ate from the Forbidden Tree,
Then planted new trees.
Modern Prometheus, you stole
The gods' fire and gave it to men.

Suspicious recluse,
You grasped the gears
Of the galactic clock.
You made the mortal immortal.

Tell me one thing that is true.

What is the measure of circle and arc?
Magical digits of *pi* –
Here, the first seven hundred and eighty six,
Sacred sum of the letters of Basmalah:

3.141592653589793238462643383279
50288419716939937510582097494
59230781640628620899862803482
34211706798214808651328230664
09384460955058223172535940812
48111745028410270193852110555
64462294895493038196442881097
66593344612847564823378678316
27120190914564856692346034861
45432664821339360726024914127
72458700660631558817488152092
96282925409171536436789259036
01133053054882046652138414695
94151160943305727036575959195
09218611738193261179310511854
07446237996274956735188575272
89122793818301194912983367336
44065664308602139494639522473
19070217986094370277053921717
29317675238467481846766940513
00056812714526356082778577134
75778960917363717872146844090
22495343014654958537105079227

6892589235420199561121290219608640344181598136297747713099605187072113499999983729780499510597105971

23

Great Newton, you hid in your rooms,
Outcast like me,
Careless of meals, stockings untied,
Drinker of rosewater, olive oil, beeswax –

You found the force
Between planets and sun,
Pattern of cosmic attraction,
Heard clearly the music of spheres.

You gauged the distance to stars
And the vast rooms of space,
Which were naught to the space of your mind.

You struck the door of the universe.
What raging night seized you
And screamed that the world
Must be number and rule?

24

Bang goes the clock in the hall,
Clamping the air, slicing the hours

Into minutes, the minutes to seconds,
Precisely the strokes of the world –
Just as the clock towers chimed
While I walked through the streets
Of Montparnasse
Foolishly wearing a fez –
Beside me my children and wife, white skinned,
Untouched by the stares.

25

I am a fragment
That hurtles through space
While the breeze of the universe
Ruffles my hair.

Evening. I gaze
Through my telescope,
Searching the colors of stars.
Some are the hues of goats' wool,
Some ochre olive,
Or pink bougainvillea.

In chasms of space
I see stars born from gases,
Great thrumming furnaces oozing their heat,
Convective motions, electron opacities –
Elsewhere stars dying,
Cold cinders
Or giant explosions, eruptions of light,

Cities consumed in a nuclear blast,
Billions of years dimmed in a second.

I have learned
That the heavens are violent and fragile
And doomed to destruction,
Just as this thimble the earth.
All in the cosmos is failing,
And nothing remains,
And we measure the hour of the stars,
As I measure one morning's light.

Here, in the glass of this eyepiece.

26

Morning. I offer to go with Abbas to the market,
He gestures me back, hobbles away with his cane,
Faster than men with good legs.

Alone, I take my warm bath, small clouds of froth,
Scent of my olive oil soap. Dressed in my work clothes
I go to the garden – the primrose,
Verbenas, the ghost flowers, globemallows,
Fragile blue bells of phacelia. Stooping on knee
In the sandy soil, pinching the deadheads,
Spent fabric falls down to the ground,
Semi-transparent, like skin.
Some of the flowers are dying and past,
But I leave them to live three days more.

27

The village magician appears in my garden,
An ancient man gnarled as an olive tree,
Wearing a filthy frayed sheet trailing
Like nightmarish bridal gown –
Holds out his hand for some coins,
And I give him white ghost flowers.

Magician, you sprinkle stray granules of iron
On your sheet, with your magnets,
And point to the intricate curves,
But there's far greater power than this.

Pattern of forces:
Magnetic that spring from electric,
Electric that flow from magnetic,
Symmetrical, right and left hand reversed,
Folded together and twisting through space
In a vibrating wave of pure energy.

Pattern of atoms:
The tiny cathedrals,
Like Chartres, Sainte Chapelle,
Framed by electrical force
And Planck's number.
Electrons that spin on their axes
Like magical tops, fixed in their energies,
Notes of a scale.
Electrons that weave in their willowy

Arcs, ordering the union of atoms
To make flawless crystals, to make
Every jot of amoebas and stars,
Each of my breaths in this instant.
And this is a thing I believe.

28

Years ago, in my studies,
I'd dive down for days
Without sleep, without food,
In a feast of ideas, books, conversation,
A thrash in my mind, how to consume the world,
Blasting and tilting life,
Singing my verses, the test tubes and flame,
Measuring movements of planets,
The diagrams, algebra, history and empire,
The ethics and dualism – all of it swallowed
In great heaving gulps, brain rush –
Intent on being chemist or poet.

29

I sit in my chair reading physics,
Old song on the radio,
Soft so it won't wake Abbas,
Sound of the sea, headdress of stars.

At the moment of midnight,
I embark on a journey inside myself,

Marco Polo of organ and vein,
My own body a miniature realm
Of the Great Kublai Khan.
What is this thing that I am,
Forces and loss?

Liver and lung, masterworks,
Hormones and nerves, the conductors
Of symphonies. Cells,
The skilled singular players –
I go deeper, am caught in a chemical sea
And electrical storms small as molecules.
Deeper, the chromosomes,
Text of my body, the ladder and rung,
Three billion steps folded thousands of times,
Going nowhere yet leading to all.
Helical molecule twisting at fifty degrees,
Constant in all living things,
Constancy fighting the inconstant stars.

30

Take this equation.
It measures the strength of a bridge
Like the bridge to Zafir's house in the sand –
Mandate with xs and ys for the load,
For the tensile strength, gravity,
Forces of molecules, torquing and strain.

    *   *   *

Purity. Capsule of number and form,
Irrefutable. Ask and it answers.

Solving equation,
A stairway of granite steps,
Square roots and integrals, tangents and sines.

I stand on the bridge –
Pure sines, derivatives
Dazzle the air, sing in their hard arcs,
Lift and parade, certain,
Impervious to boom and blast,
Concrete from concrete

And mountain from number,
Its sureness, my sureness.

31

My birth no one remembers –
The cries of my mother, my father
Away for his business,
No record of who first said "God" to me,
Who rubbed the date on my mouth,
Pattern of light through arched windows,
The scents in the air – uncles and aunts
Have no memory.

But the birth of the cosmos – the heat
And the densities follow equations,
The temperature 1.5×10^{10} degrees at 1 second.
Twelve billion years
In the past, all began:
Time formed from nothingness,
Space formed from nothingness,
Universe born by some chance, or not,
Moment of ripe probabilities,
One of uncountable eggs.
And then the explosion,
The energies opening space
And the boiling of light
In the Bang without eye without ear.

    *   *   *

And this too a cosmos of endings
At billions of years in the future
As gauged by the mass and the speed
And the clock of the flight of the galaxies.
Billions of years in the future,
The galaxies, caught in the ocean of space,
Sail apart as the cosmos expands
And the density dwindles to nothing.
The stars spend their energies,
Light fades and dims,
And the galaxies ghost ships
Adrift on an infinite sea,
With no heat, with no life.
But the numbers remain.

32

Mother would come to me late in the night,
After her concert,
Humming some song she had sung –
Hair down and barefoot,
Her fingertips brushing my cheek
Smelling of jasmine and sandalwood.
Hers was the breath that could flow in the night
Through the hallways with flickering light,
Voice that said, "Certainty."

She forgave everything –
Money I squandered,
My fidgeting poetry, years living abroad,

Wedding a foreigner,
Children she never knew,
Return in despair –
Loathing my father,
Even that she forgave.

And she sat with me during the night,
Stroking my brow, while I listened
To waves of the sea –
You are the question and I am the answer,
So breathe with me.
I am the sea that rolls over you,
I am the green and your comfort,
Say yes, then say yes, then say yes.

33

Out of the mist, two birds emerge,
And they glide without sound to the sea
In a wraith sweep of dream wings.

Two hawks from the mist.
I can see them, just past
The stone steps and wall,
Sweeping in slow ornament.
Uncover my head, let the wet
Damp my face. I receive all
That comes down from the sky.

34

Bits of my life sideways in drawers,
In the cramp of my closet,
The dog ears of books, odors
That never dissolve, photos
I cannot behold, cracked spool and string
That was once my son's yo-yo –
He touched it, then me.

With this string and this weight
I will compute the Earth,
I will fashion a pendulum.

Release the weight,
Set the bob swinging.
It follows the arc of the world,
Back and forth, back and forth,
Beating like beats of the blood.

I can measure the time of each swing,
Record the number of seconds
To tenths and to hundredths.
The number, I write in my book.
Now, with the laws of mechanics and force,
By dividing the length of the string
By the square of the time
I have found the Earth's gravity, 9.8 meters
Per second per second.
A number that holds other numbers:

32

The mass and the size of the Earth,
Figured with only a clock and some string.

These are the things that I know:
Heart that repeats like a pendulum,
Muscles that slacken with time,
String that he touched and then me.

These are the things I want to be true:
Seconds to measure a movement,
And movements to follow a movement,
Each action preceding reaction,
Like heat and the wheeze of Abbas' kettle,
And hope that might grow from reduction,
And knowing that comes from experiment,
Nothing that hinges on chance,
Even the fall of a pin,
Cause and effect –
Push and the Earth pushes back
Just as it must, hundredths of seconds, exactly.

35

This is the cosmos of measuring.

Numbered the barrels of olives,
The crates of pocked oranges,
The kilos magnesium sulfate, the times
Of the shippings, the paychecks, the days

Without rain – all of it kept
In these columns and rows.

Great Lavoisier, bony nosed,
Eyes far apart, silent,
Preferring your textbooks to play –
You were the master of measuring.
You were the ruthless accountant
Of chemical quantities,
Reckoning mass never vanished
Despite changing form.

With your scale and your balance,
Your cylinders, candles, and bellows,
You tallied the weights of the world.
Phosphorous, sulfur, and minium
All breathed your air.
You found sharp truths in the numbers and grams,
Conceived substance as change,
Transformation, with quantity always the same.
You found the secret between the old elements –
Fire and water and air – joined by oxygen.
You believed only in what you could measure:
"Trust nothing but facts."

You were Magellan of molecules,
Substance, and flame.
You showed the cleaving of food
In the body to be a cold fire.

        *   *   *

34

Great Lavoisier, you counted grains
Of the vast hour glass
In a slow summing up of eternity.

36

Lavoisier –
Can you compute
The Patio de los Leones,
The mosque of Alhambra?
We'll number the columns
That plunge from the arches
That lead to the heart at the center,
Stone fountain of lions –
We'll measure the tiles and mosaics
Adorning the walls of Dos Hermanas,
Rich mathematical patterns
And infinite regress to truth,
Study the falling light,
Shadow and maze,
Gold covered filigree, each
Line a part of the cosmos
Of number.

Dust of the lions has smeared on my hand,
Darkened my tunic, my bed sheets,
My fingerprints smudged on my ledgers,
More waste in my wasted house.
Can this too be measured and weighed?

Down to each molecule, atom,
And even within, to the energies,
And then to the spaces of nothingness?
What is the number of nothingness?
Wait – I will count.

37

I hear voices.
Abbas' grandchildren have arrived –

Zarina, with two missing teeth,
Tabat with one skinned knee, dark eyed Sabrine,
And the naughty Ra'oof.
They play jigsaws while eating sweet cakes
Soaked in honey and filled with moist dates.

One by one, they come to me,
Calling me Uncle and kissing my cheek,
Each kiss a sting of remembrance.

They follow Abbas to the groves –
"Help this old goat" –
Plucking up low lying oranges for balls,
Pitching and zig-zag to sea, splashing
Romp, trail of wet sand and shells,
While Abbas on the terrace is bellowing.

    *   *   *

What would I say to my own children and grandchildren?
How would I listen?

38

Night, and the children have gone.
Abbas snores on his cot.

The Voiceless has come and he waits
At my door. He has traveled from deserts,
Has sand in his shoes. Mouth
With the motionless lips and the question
That cannot be asked. Standing, he waits
In the night, in the dark, and he smiles
With a terrible grin. From my window,
I look at him, dim in the lamp light,
And dimly he grins at me.

Caught in our mutual stare,
He throws stones at my house,
Rolls his eyes back and forth,
Gestures to open my door.
I refuse.

He continues to grin,
Hours pass.
With the dawning, the Voiceless departs,
And returns to the desert, his unspeaking
Mouth, but he leaves a dark mist

In the bloom of my house, he leaves
Sand at the base of my door.

39

I ask: What is the form
Of the principal forces, the atoms of atoms,
The dark energy pushing the galaxies?
How does the smell of baklava
Fix in my brain?

Numbers and names I will give to these things:
Gravity, chromosomes,
Seconds, and quarks. Photosynthesis,
Synapses, covalent bonds. Mitochondria, quanta,
And thermodynamics. Electricity, amperage,
Codons, osmosis, cortex, and catalysis.

Is a thing known if it cannot be named?
There, on my terrace,
Between those two stones –
Flower whose petals are shapeless
And odorless, empty of color –
Unnamed though each atom is trapped.

40

Abbas and I shovel sheep dung,
My olive trees hunched
Like a throng of old men –

Sun overhead, I look up
And I gaze at the clouds –

This is the way that I see:
Ten trillion photons of light make their way
Through my pupils each second of time,
Through the oval-shaped lens, through
The jelly-like fill to the retina, hundreds
Of millions of cells, where each photon
Of light meets a molecule, retinene, coaxing
It straight from its twisty vine,
Curled bougainvillea-like. Neurons
Respond to the dance. Protein molecules shift
In their shape, so that sodium cannot find

Passage, electrical charge unrelieved,
Shudder of current moves through
The moist neurons and flies to the folds of my brain.
Here the fourth layer is sentry, receives
The first tingle, computes a sensation,
And passes its tremblings to five other sheets.
In shards of a second, some hundreds
Of millions of neurons start quivering,
Each shocked by one thousand
And doing the same to one thousand more.
Click-click-click sound the firings,
Some punctured and scattered, some synchronized.
Click-click the pulsings in waves –
And my brain tells me "cloud."
I believe in the knowledge of sight.

41

Newton and Darwin, Pasteur and Al-Haytham, Mendel,
Mendeleyev, Curie, and Bose. Galileo, Bernard,
Lavoisier, Al-Biruni. Einstein and Watson and Franklin
And Crick. Here are your robes and your sacraments.
Here light the lamp of eternal oil.

Kepler and Brahe, Berzelius, Dalton,
Copernicus, Boltzmann, and Bohr.
Pauling, Boveri, Planck, and Cajal,
Heisenberg, Meitner, Brown, Krebs, and Schwann.
You burn the incense of asking and knowing,

Toss petals of restlessness.
Is nothing still?

42

This is a cosmos of living things.

Darwin – sailor, surveyor,
Collector of bottles and crates,
Casks crammed with pickled fish,
Bird skins, dried plants –

What beasts you'd find here:
Camels, gazelles, jackals, hyenas,
Sleeved mouflons, horned vipers and cobras,
The bitterns and egrets,
The rails, crakes, and coots.

Cutting blue eyes, bushy brows,
Clumsy of movement and awkward
Of hands, with a passion
For birds and rare bugs.
You grasped the rule of survival and change,
Found the origins: petal and pistil,
Proboscis and lung, myriad shapes
Of the beak. Three kinds of mockingbirds,
Each from a new island. Extinctions of capybara,
Sloth, armadillo. The flightless small rhea
In south Patagonia. Lizards.

And tortoises, different by size of their shells,
Island to island.

You showed that
Order can grow from disorder,
And purpose from aimlessness.

You said: "The large hand of chance is the hand
That has fashioned the cosmos of life."

43

World of unending forms,
Spun from one primeval pattern?
And clouds,
And the infinite turnings of shells?

You found the links
In the hard chain of life.
You found that many can unfold from one,
Found the complex in unity.

Walk with me, Darwin, to prune
In my groves, speak of the crush to survive.
Goats in the stone keep
And gulls on the shore, falling
From ancient beginnings, with blood
Of our blood, kin in our family,
The endless unwinding of common

Beginnings, the branching and branching
Through layers of time,
From the first tiny movements
Of life on this world. I am goat, I am gull,
I am part of their bodies. Each breath
Is the breath of millennia.

44

Afternoon on my terrace, I sit
In the heat, bare chested,
Sipping hot tea. Far off,
The sound of an oud. And then nothing.
The air closes, silence. Again,
The faint strings of an oud,
Muffled bells, drums,
Horn sound of mizmar –
All swallowed in air.

In the distance, I see figures.
Dim movements emerge from the haze,
Women with parcels on top of their heads,
Men carrying carpets and tents, drum players,
Trio of mizmar, boys trailing bottles on string –
It's a wedding procession, defiant with music,
Making its slow way up the coast.

The bride, covered rich brown in henna,
Walks with her eyes on the ground,

The Koran lifted over her head,
While the mother meanders from guest to guest
Pressing a coin to each brow.
Scarlet veils billow like sails
Of a boat sailing on sand.
Servants bear baskets of lamb, peppers,
Chickpeas and couscous, zucchini,
Cinnamon, tabil and fish,
Ladoos and coconut cakes.

They are so young.

And then they have passed, winding north,
Fainter and fainter, a flicker of color,
The white dot of a caftan, dim note
Of one horn, and then gone.

45

Unlike Zafir's second wedding,
A gluttonous day in his dry desert house,
He just turned eighty, his bride twenty-one –
Oud players dressed up like ouds,
Tambourines bulging with lamb,
Stuffed brawled and slammed –
I on the sofa asleep
When Zafir, bloat-bellied, places my hand
On his young bride's gold bracelet,
Says: "This will endure beyond all mortal life.

Gold. Density 19.3, mass 197. Kiss it,
You've kissed what is true."

46

To know of this world,
Should one not love details?
Here, the DNA codes for the essences of life:

GCT $\longrightarrow$ Alanine
CGT $\longrightarrow$ Arginine
AAT $\longrightarrow$ Asparagine
GAC $\longrightarrow$ Aspartic acid
TGT $\longrightarrow$ Cysteine
CAG $\longrightarrow$ Glutamine
GAA $\longrightarrow$ Glutamic acid
GGT $\longrightarrow$ Glycine
CAT $\longrightarrow$ Histidine
ATT $\longrightarrow$ Isoleucine
CTT $\longrightarrow$ Leucine
AAG $\longrightarrow$ Lysine
ATG $\longrightarrow$ Methionine
CTG $\longrightarrow$ Phenylalanine
CCG $\longrightarrow$ Proline
AGT $\longrightarrow$ Serine
ACT $\longrightarrow$ Threonine
TGG $\longrightarrow$ Tryptophan
TAC $\longrightarrow$ Tyrosine
GTA $\longrightarrow$ Valine

47

Journey of questions, I paddle
My oar in the stars.

I look out my crumbling arched window
Across the orange groves –

There, in the dusk, the great tower
Beckons again.

I return to the infinite hallway –
Tell me: What is the center?

The great door swings on its hinges,
The hall gyrates and twists, I grow dizzy,
My seconds churn hours and seconds again.
"You are asking of time and of space,"
Speaks the universe. "That is the center."

Great Einstein, you were the master
Of time and of space.
You cracked the clocks of the universe,
Fracturing glass and coiled springs,
Showing that seconds and meters
Are not what they seem –
Time does not flow at a uniform rate,
Strumming and sliding through pages of space –
Seconds, like rubber, can stretch
And contract. Moving clocks won't keep

Their synchrony, caught in a monkeydance.
How did this gangly time pop
From your symmetries?

48

This is the world of the ticking of clocks,
Menses of women and tides
Of the moon. Orbits of planets,
The swing of a pendulum, spin of the earth.

Here, at my table, I question time's meaning,
I gaze at the legions of people who pass by my gate
And ask: Where are they going, and why?

Two hawks alight on the rail with a flap
Of dark wings. Are they time and not-time?
They watch as the throngings of travelers
Pass silent below, the successions
Of parents and children, the deaths after births
After deaths through the sweating and splashing of time,
Pendulum's swing and the next and the next,
With the endless repeat of forgotten lives.

What is this passage of seconds and centuries?
Cyclings of atoms through mindless
Vibrations, this flight of the galaxies
Racing to where? What meaning this instant
Of time with my inhale and exhale, this moment
Of breath in infinity?

49

Einstein, moustached and sad eyed,
Can you hear the blind ticking
Of clocks? Can you feel each soft second that slips
Through the glass? You were young at one time.
Can you make the world young? Can you
Rebuild the world at each dot of time?

Exiler of absolutes. Nothing is still
In your universe. All speed is relative –
Speed of my walking astride the chipped stone,
Speed of this planet bound to this star,
Speed of my breathing through wind,
Speed of my thinking compared to the absence
Of thought, which alone is dead center,
The motionless point.

Einstein, philosopher poet,
Iambic geometer –
Dreamer of symmetries, perfect
Reflections. Proclaimed that "The world
Must be this way and this." And the
Universe answered "Yes."

Tell me one thing that is true.

50

Einstein, were you stunned
When you found that the hardness of time

Was illusion? Gossamer.
What else did you doubt? Matter and shapes?
Fact of your birth? Lung in your body invisible?
Waking or sleeping, how would you know?
Did you protect the surprise of your mind?

51

This is the cosmos of time and of space,
And of light rays that travel twelve billion years,
And the whale raptured sprawl of the galaxies.

But is this not also the cosmos of life,
That rare cluster of atoms and forms,
A few grains on the beach of nonlife?

Is this not also the place where I breathe,
Where Abbas' cooks his lamb and tajine,
Saffron and cardamon seeds?

I knock on the door of the cosmos of life
And ask: What is the reason I breathe?
I'm answered: You breathe to make energy.
I ask: What is energy?
I'm answered: The movement and heat of your body.
I ask: What is the measure of movement?
I'm answered: A change of place over time.
I ask: What is the nature of place and of time?
I'm answered: You return to the center.
I ask: What is the nature of thinking?

I'm answered: The spasm of cells in your brain.
I ask: What is the thing that makes spasms?
I'm answered: The movement of positive particles.
I ask: What is the nature of movement?
I'm answered: You return to the center.

One thousand questions, and each gives
An answer, which then forms a question.
The questions and answers will meld with each other
Like colors of light,
Like the light rays that once crossed the space
Of the cosmos
And rest now in the small warmth of a hand.

52

I launch my small skiff in the sea,
Water still cool from the night,
Oarlocks corroded with salt rust.
I row for a jut up the coast,
Quivering ocher in sun.

Then the sea thickens, my oars
Become mired. I stare in the water
And see my own body
Just under the surface and clinging
With arms to my oars. "I'm the end of you,"
Whispers this other me. "I perceive all."

      *   *   *

I try to shake off this dark clinging thing.
I'm unable. It heaves on my oars,
Trying to wrestle me out of the boat.
"I'm the end of you. I know the day
Of your dying, your silence, your pleasures,
Your sufferings, when the last night
Will swoop down on your world."
I feel my mind bleeding.
I strike the thing hard,
And again, and again, and at last
It surrenders the struggle,
Releases its grip, slowly sinks and submerges,
Pale face that grows dim and is gone.

Part II

Questions without Answers

53

Late afternoon, and I rock
In my chair, breeze
Roughs the olive trees,
Jangles the chimes on the wall.

Little paw brushes my cheek,
Lightest of touches that comes
On the wind, a familiar sensation
That moves to my belly and stays,
Plays like the moan of a sad jazzing horn

52

In the small hours of morning
When everyone's gone.
This sensation I know –
Why has it touched me this moment?
I strain to remember, but all I can find
Is this moaning –
It grows as it darkens
And spreads through my body.
It empties me.
Is it the song that my mother once sang?
Or the sound of the car waiting in rain
On that last windy night?

And then it is gone, hushed as it came,
Shapeless, unmeasured, unmeasurable,
Leaving me still in the late afternoon.

54

In the mirror, I gaze at the road
Rushing past and behind,
Hurl myself over the sand
While the desert unfolds like a roll
Of brown paper, air breaks
With the swoop of the bustards.
I want to remember
The small crusty town and the prefect
Who yammered and pinched his bent nose,
Smell of cooked lamb in the dusk,
Reason I went there.

Perhaps I was searching for comrades,
Or cousins or uncles who knew me
Before all those years disappeared.

But here I am, middle aged,
Scraggled and roaring the desert
To some other small town –
Or maybe I'm just in my chair,
Musing on fast cars and wine,
Hours of my youth,
Staring at shapes in the sea.

"I'll tell you a story of old Zafir,"
Floats the voice of Abbas –
What? Have I been napping?

54

I try to stand, staring
At shapes in the sea,
Head lolls in afternoon heat,
Humid air rolls through a window –
"Zafir once lost all of his money,
Bad loans to a cousin in Tripoli . . ."

55

Restless sleep, musical shadows
Slip sidelong across my bed -
Is it mirage in the night, light
Of the infinite hallway, sand serpent,
The place of locked rooms? I knock
On door after door, and I listen for footsteps,
But I am alone.

In the darkness, the silence, faint drumbeats
Sweep under the doors, tympani,
Kettles so slight that they
Might be imagined, now here and now gone –
Tap disha tap -

Where are my olive groves,
Puyas and poppies,
My primrose and red bougainvillea?

I peer down the hallway.
A gauzy light oozes
From under the doors. I touch each

Of them – intricate carvings
Like faces explored by my hands.
Unease flows through me,
The darkness, the waiting, the unending
Hall of locked doors.

56

I know the masses of manganese, gold –
But why is there gold at all?
Planets and flesh?

I listen, but all I can hear
Are the faint rolls of drums.

I call out to Newton, al-Haytham,
And Darwin. I call out to Uncle Zafir.
But I am alone.

I wait at the doors – and I notice:
A finch flies to the window,
Alights for no reason. The bird begins chirping
In weak syncopation
To sounds of the drums.
Sings and then quickly is gone.
Was it a finch or a warbler,
Or no bird at all?
Clouds drift through space, mask the sun,
The light darkens, then blossoms again.

56

Time passes. Where am I?
There on the tiles of the floor, something,
A paper scrap crumpled.
I bend down to read it,
Some writing that's smeared –
Nonsensical scribble, or message, or prayer?

57

Is this the cosmos of answerless questions?
A world of unreadable writing?

A cosmos of good who do evil
And evil who do good –
And the question of Hamlet
While ranting his tower for days
As I rant in my silence
And wander the rooms of this house –
Is this the world of the ghosts of our fathers?

Where is the knife that can cut?

Here, let my psyche be my temple,
My breathing my candle.
My sacraments stories and songs,
Tympani, cloud shapes and paintings,
Some whisperings, letters that wait
To be written, and words
Of philosophers, romance and illness,

My own death approaching.
And still I am blind.

This is a world not of action reaction –
But each action questioned forever,
Where lust defeats virtue,
Where motives descend
In a watery bog –
Things and their opposites both can be true.
Was that not the thought of Lao-Tsu?

Can I give up myself
To this desert of night?
Witness again what I've done
And not done?

58

I knock on the doors of the universe,
Asking: What makes the swirl
Of ghazali love songs?
And the parallel singing of loss?
And the choice to live life alone?

I surrender my calipers, rulers, and clocks,
Microscopes, diodes, transistors,
Glass flasks. For how can I measure
The warmth of a hand? Or dissect a face
With the digits of *pi*?

* * *

Thus, I stand naked, with nothing
Except a fierce hunger to fathom this world,
To embark on this road
Without length without breadth.

59

Dawn – my calipers cracked
On the walkway,
I sit on my terrace, drink tea
From a porcelain cup,
Watch a sun touching sea,
Watch a fire created by water.

And I dream of one thousand suns
Caught in one thousand seas,
Slowly dripping their way to full blaze
In some galaxy flung through the wasteland of space.
A thousand new dawns for what purpose?
The stars and the comets, the masses and forces,
The gold drops of tea in my cup –
Is it all unintended? One blithered yawn?
Does the universe know of its spinnings
And forms? Divide flesh from the fleshless?

As I sit on my terrace, I wonder
If something can see me, some
Vast lidless eye. Or perhaps hear me,
My murmurs and breaths.

    *   *   *

Softly I utter from Omar Khayyam:
"Into this universe,
Why never knowing, nor whence,
Like the water, willy and nilly flowing."

60

I whisper the name Mother gave me,
A private note passed in the dark,
Here, off the hallway,
This sitting room swimming in light,
Primroses outside the window,
The orange trees beyond, that's where
She said it – then to her piano,
And I to my school books, my sea glass
And bottles – still here, in the room
With arched windows,
The chipped russet tiles, rotting damp smell,
Textbooks still here,
Pieces of glass that I picked years ago,
Rage of the blossoming sea, and I
In my boyhood, unknowing of wine
Slightly sipped in the Christian cafes,
Of the women who look men in the eye,
Teeming sad nights, and my own crashing winds,
Flung to the north, and the booming years –
All of it lost, leaving only this broken
And empty house, Abbas here taking his nap,
Ghosts of a family.
I would have been chemist or poet.

61

Ghosts of a family. There she is middle aged,
Age of me now, sewing a shawl in her room.
I have returned. Mother, please sing to me.
Yes, and her voice is the voice I recall,
Fragile, a flight of small birds,
But she won't speak the name
That she gave me in childhood, our secret.
She touches my scars with stiff hands,
Fingers that once touched the keys,
Holds me, but she will not say the name.
Years turn to years, Father long dead, I in my room,
Mother in hers. She grows dimmer and dimmer
Until she is gone.

62

Dusk on my terrace – a man wanders by,
Plucking his oud, wearing billowy bisht
And a fez with a red colored tassel,
With music that dances the air.

Says he'd been walking eight months,
Town to town up the coast,
Seeking the wife who disdained him
And wanting her back. Says
That in each pair of strings is a sadness
And joy. He seems glad of his torture.

    *   *   *

All of it sung to me playing his oud,
With its pear shape and bowl.
And I offer him olives and eggplant,
But he will take none.

Then the darkness –
He drifts away.
All I can see is his teal sash,
Swaying and small,
But his oud can be heard through the night.

63

Haze drifts from the shore,
Softness of air, and a veil drapes the sky.
I open my window and gaze at the sea.
I imagine boats sailing,

The cabins of sailors, the islands,
The cities, whole continents lost
In the mist. And I wonder what parts
Of my body lie outside my self:
Lungs and the liver, performing
In secrecy, kidneys and spleen
And their intricate passageways,
Miniature battles engaged
By my cells in their silence –
The chemical messages
Sent in the dark, thousands
Of impulses launched every second,
Dispatched and received
In unconsciousness.

Throbbings of worlds in my body, unseen
By me, unaware – What thing am I?
Many or one?
Where is the nub of me?
Is it my spongy gray fold of a brain,
Nerve endings,
Lopsided cavities?
Which is the piece that picked glass from the sea?
Which is the piece that conceded to love?
And the piece that spent love for this home
That is homeless, spent blood flesh
For bloodlessness?

From where comes this feeling of wholeness
When so much is scattered, invisible, mute?

64

Lao-Tsu, Rembrandt, and Omar Khayyam –
Be my companions on this dizzy journey,
This twisting and thrashing to nowhere,
Reversals and formlessness,
Numberlessness, pageant of nothingness,
Strivings, nonstrivings,
And ends from beginnings,
Beginnings from ends
And the unsmiling smile
And the swing of the clock
That has never kept time and the moon
That is whitened and shale,
Cities that wither to dust,
Heart that both loves and disdains,
Guilt of the guiltless and guilty.

Must I surrender the line and the rule?
Reason and number have failed in their triumph
And triumphed in failure –
But how can I know, for there's nothing
I know. And yet here I am searching. I vibrate,
I shudder with life, roar and balloon
In this evening of primrose and morning world.

65

Footsteps and thump, Abbas is back
From the market with sugar and coffee,

Black pepper and caraway, guram masala –
His face blotched from sneezing.
Smiles at me, chewing some seeds.
"I'll cook a tajine," he says, wiping his nose –
Chick peas and lamb, sun dried tomatoes,
Fresh cheeses and eggs, onions, saffron –
"That wild boy on the hill won't pay for our olives."

"Let him have a few barrels," I say.

Abbas laughs. "You're not like your dad.
Have you seen our accounts?"
Shrugs. "No matter to me, I'm an old man."
Then quoting Khayyam:
"Thou art but what thou shalt be –
Nothing – thou shalt not be less."
Laughs again.

Goes to the sink for his *wudu*,
Washes his face, ears and mouth,
Nostrils and hands, arms to the elbows, muttering prayers.
"At least let us be clean."

66

How does it start? Where does
One find the beginning? Not now
With these wakings, but then, years ago,
When she spoke while anointing the rose.

  * * *

Who could imagine all this: cities and towns,
Fruit stalls, the bowls carved from trees,
Gowns of the brides,
Words that she spoke as her lips
Touched the rose.

Can I walk back through mansions of time,
To the moment she first saw me,
Quaking and heaved in my youth,
Dark skinned and flung on her shore?
Or before, when I squirmed in the wealth of my father,
While she – continents distant, oblivious –
Slept in sweet honey?
Or was it before, in our parents – was there a sign?
In the details of two countries?
In seasides and rooms
Where decisions were made?
Was there no note of our future,
Our flamed cataclysm?
What sounds could be heard in past centuries,
Primitive air?

And before. In the time before time
And the space before space. In the wavering
Haze of infinities. Does the universe know
Of its future unraveling? Was there no hint
Of her lips on the rose? And our meeting,
Our union, the births, and the lives?
This planet, this second, the dress that she wore,
Red of her lips touching red petaled rose –

Was there no speck?
Yet her lips touched the rose. There, I can see it,
The silk and her hair.

Perhaps I might hear some faint echo –
I listen to scrapings and breath.
Is it Abbas now asleep in his room,
Or the wallowing mouth of the sea?

67

When was the moment?
The placement of chairs,
The duration of glances,
The scents in the air?

The tinkling of glass I remember, the wine
And the chattering, people in motion,
A flute solo, tattooed guitarist, autumnal
Breeze from the window, the winding
Of tendrils of plants in the sill.
Foods of a taste strange to my tongue,
Pâté and brie,
That, I remember. But when was the moment?
The people in motion – there, at the window,
I saw her. Was that when it happened?
Or later that night, on the shore,
When all we could see were the dark silhouettes
And the boats tied with rope to the pier?
Or weeks later, when I only imagined

I saw her, imagined the shape of her lips?
Or perhaps it did not happen at once,
But instead slowly – parts of me stolen,
A piece, then a piece,
Until I did not know my right hand.
Ravaged invisibly,
Seized without speaking a word, wanting it.

And the demon who lived in the water,
My other self, grinned the most horrible grin.

68

Demon? My long missing wife?
Have I dozed at my desk?
Afternoon swoons in the heat, waking and sleeping,
My neck moist and sticky with sweat –
What is the hour?
These deserts and shambling old orchards,
These orange groves and goats,
Villa, this room and this desk, splintered chair,
Smell of my own fingers that lift up this pen –
Here. By some miracle, I have awakened.
Today. Why, I don't know.

69

Cosmos of formlessness, tell me a tale.
Tell me a story that might have been true –

What if I'd stayed in that country
Across the sea? Waited each morning
For shouts in the streets, knocking on doors,
Clinking of bottles and glass,
Always the need to wear strangling shoes,
Wedged in a house without gardens,
Cold river, my blessed two children and wife?
Could it have been?

Let me hear stories
Repeated by voice through the ages
And changed with each telling, like wrinkles
Of water in wind, like light cast away
From a bowl of brown dates.
Some are my story.

Tell me the words that have never been stilled
By internment in books, the rebirthings,
The stories with endings unknown by the teller,
The whisperings of bedouins who camp
In the night, and the howlings of sailors
In ships on the sea, and the voices of lovers
Who meet in the darkness
To name without names.

I give up the quiet.
I give up the constancy.
Here are my ribbons of skin and my pieces
Of heart. Here are my chunks of a brain.
Retell my life.

70

Years before, when it started,
I was on fire –
Breathing pitched
Belly contorted in anxious folds
Wanting and gimpy thighs
Pleasure and pain, and I screamed
For the burning to finish, but I did not wish it
To end. I would drink that sweet fire
And watch my destruction and flame.

Was that the desire that doomed me?
Or was it my fear?
I still burn
While the sound of her voice echoes
In vast burning canyons, her fingers draw circles
On burning skin, nipples set fire
To my chest. That was the moment,
I could not extinguish the fire, I could not remember
My journey across the sea,
But only that instant
When I was ablaze
With a sun in my body,
A perishing star
In this perishing universe.

71

All who have travelled this perishing life,
Let us gather and wait for our healing.

But time is no healer,
And time too will die in the vanishing stars.

Great Rembrandt, the master of light and of shadow,
Of tortuous path, ambiguity,
Come paint our faces,
The dazed lakes of eyes wishing for some
Other life, jowls full with unfinished living,
And brows soft with unceasing hope -

Come paint our faces, the cradles
Of sun through white shutters,
The graveyards of dark afternoons,
Stirrings of tea in a lifetime of mornings,
The touch of the lips kissing skin –
Yes we remember –
The plantings of seed pods that may never bloom,
Visits of uncles, the births of our children:
We've witnessed it all, without knowing why.

Come paint our faces,
The lights and the shadows,
The ends and beginnings,
All lost in the sea of uncertainties.

72

Here is the globe on my shoulders,
The folded gray clog of a brain
Whispering incessantly, needless of lips,
Multiple voices and dissonant,

Parsing the world into thousands of choices:
To say these few words or some others,
To steal what is waiting or not,
Give or withhold,
Action inaction,
Betrayal or loyalty,
Kneel or stay standing,
Live or decide not to live –
Bodiless voices, each states its case
In the bodiless court in the room of my mind.

In the wood paneled chamber,
The waist-high bar with the smell of tung oil,
The invisible barristers each claim to be me.
"Do this," says a voice, and "Do that," says another.
But I did not ask
For this splintering of selves.
I want only to sit at my desk
Looking out at the white spew of sea,
Musing on what could have been.
My head will explode, flinging
The parts of me, clashing
With full ladled screams.

Here is the globe on my shoulders,
The folded gray clog of a brain –
I will pretend I am one, only one.

I am the one who once gathered the glass,
I am the one who took girls to the sea,

72

I am the one who could not sleep in the night,
I am the one who lost life to life.

Outside the room of my mind
Someone speaks to me – I will speak back,
While unseen in the chamber,
The barristers quarrel and snicker
At this grand faade.

What – have I uttered a word?
I'd expected to babble in disjointed sentences,
Stepping first left and then right.

There is no end to it, this multiplicity,
Life in this throng of myself,
Self an illusion, this oneness illusion.
I thirst and a thousand me open our mouths.

73

Dryness, this heat,
I am soaked with my sweat,
Napping and dazed –
I hear voices outside of my window,
A gang of men dragging large sacks,
White tunics limp in the sun –
Traders of coffee, their teeth brown
From sucking on beans.

        *   *   *

One man I know,
Thick fleshed, perspiring.
I call to him, mentioning meals that we shared –
But he stares at me cow-like,
He does not remember,
Says I am confused.

Says that he's traveled to Laos, to India,
China, New Zealand, Azores, has labored
As merchant and carpenter, sailor, spice grinder –
His hands are the color of cinnamon.
"One place, another," he says. "None of it matters,
It all ends the same."

74

Day drips to night
Becomes day again night,
Hours slowly pass without sound –

As I sleep in my bed smeared with cumin
The night dancer spins in my room –
Or was she there always, that dark in the darkness?
And have I now found her,
The whispering voice of the doors?
And she moves to the rhythms of drums,
Slippered toes touching the floor
With a sound like sails ruffling in wind.

Night dancer, please.
Let me show you the book of my life.
Here are the beards of my
Grandfathers, letters exchanged
By my parents, the moment of sun
In the groves – yes, I remember –
The songs and the schooling, the small
Dimpled pot, sorrel rug stitched in the corners,
The lists of my vanities,
Youth drunk with restlessness,
Poetry, typescripts, inventions,
Reports and bank statements, the notes
From the dealers, the names of my lovers,
French wife with her turquoise eyes, children,
The land that despised me, my otherness,

Cruel looks in the street –
Longing for orange groves and sand,
For the primrose and aloe, for saffron tajine,
For this sun varnished villa that sits on the sea,
For my sea glass and room –
Then my escape, flagrant abandonment,
Shame. It's all here, I can certify.

Can you make dance from it?
Let me help. Tear out the bank statements,
Diary of lovers, reports from the schools,
Graphs and equations – surely none lyrical.
Compress the rest small as a grace note,
Lighter than breath. See how it floats.
There, it has slipped
Through the window, so weightless,
Like me, mingling with millions of particles,
Lost in the ocean of air.
Did it make one arabesque?

75

Dancer of night,
Did you read of my sins?
Speak to me: Why is there sin?

And she flutters in dark soubresauts.
"Each thing exists with its opposite,"
She sings, singing Lao-Tsu.
"Pain lives with pleasure,

And illness with health,
Evil with good,
Force with the absence of force,
Motion with stillness,
And being with nonbeing."

I dream in my spice scented bed
While the night dancer
Turns in the room, moving and still,
Time and nontime,
Awake and asleep.

76

I wake in the night – snores of Abbas,
Ticks of the clock, tinkle of bells –
There, out the window, the bubble
And oil of the moon, dripping on steps
To the terrace, on split bark and leaf,
Moon on my boat beached on the shore,
Whitened against the black sea.

What do I want in this hour?
Should I put on my sandals and walk in the dunes?
Should I read, should I eat,
Spangled, confused?
So much is hidden,
And all doomed to waste away.

    * * *

Hour by hour the taut wire slackens,
The sharp pull of memory, air and regret.
Something takes hold,
Grinning that terrible grin.

What will become of my flesh
When I've passed on to nothingness?
Atoms, my atoms, still bound
By the forces of physics
But free from their duties to life –
Where will they travel? Forgotten,
The sight of an island in haze, the explosion
Of starlings from trees, squeak of a door,
Darkness and light,
The procession of seasons –
Forgotten and lost from the bits of me,
Scattered in soil and in wind,
Drawn through the roots of a Chinese hibiscus
To sleep in its white sleeping petals.
Some washed to sea, swallowed by fluke
And blue marlin, some tossed in the air,
Inhaled and exhaled for centuries. Strangers
Unborn will then breathe in my atoms,
All of them there –
Lavoisier can count.

Where is the moment I wept for my mother?
Last sight of my children? The memory
Of poems and geometry?
Atom by atom still there, but not there.

Minnah –
I still have this ribbon
You wore in your hair.

Do you have children?
And passion, and sorrow?
I'd cut off my hand
For a glimpse of you.
Can you forgive?

Did your mother not tell you
I asked that you all come with me?

I wish you had known of my agony,
Caught in the vise of two deaths.

Hassan –
You were so small,
Brown-eyed and quick as a fox,
Laugh like your mother.
Can you remember me, anything,
Walks by the river, my voice?

77

One hundred mouths chew at me,
Saw through my bristle and skin,
Swallow and dribble blood,

Gnaw down to bone –
Still, I will not be clean.

How can I kneel to God,
Blackened and fouled? Faithless,
For God has forsaken me,
Sanctioned it all,
Knowing the words spoken,
The movement of feet, fire and death.
I could throw my whole length
Across his fierce mouth,
And he would not utter a word,
Not ever save me from this flaming world.
Where is his power? It fades
Like my daughter's dull ribbon,
Outshone by my sin.

79

Perhaps there is more than one,
Multiple Gods.
Do they vie for each choice
By my unwitting brain, each sliver
Of light of auroras?
Or do they remain in their separate domains,
Each taking his own feeble steps?

Or is there no God in the multiple worlds?
No mind that says: "This is the way it will be.
Here are the continents, here is what moves."

Without God, I would watch as the moon bends
Through night, slaving to gravity. I would say only:
"Now I am here at this moment.
My life is my own."

God of all Gods or none?
Something or nothing?
Mind or a mindlessness?
Speak, I will hear you, I'm waiting,
I give you the knife,
Give you my flesh.
Open me,
Show me one vein
In this vanishing world.

80

Can I remember?
The glances, the wind, angles of walls?

There, in the mirror, I see
Marks that were not there before,
Minuscule shiftings, indentions,
The skin of new crevices,
Baldness and moles,
Eyes with their alien stare.

Down to my mind –
How can I know who it is?
Moment by moment, I feel something shift,

Mind that re-minds after each spoken word,
Each action taken,
Recrossing filaments, neurons,
New pathways joined, old bridges broken -
Brain of perpetual nowness.

I want back my mind.
But I ask: Who is it wanting?
Who was it slept in this body last night,
Stood on the terrace as windy sand
Pelted the blue stucco walls?

There, I can see my new face in the mirror.
I raise up my hand,
And a hand rises up at my side.

81

Now, I remember that small sandy town
On the edge of the desert –
The prefect, degrees on his wall, medals,
And children, drove off one day
For no reason and never came back,
Chasing some wild something,
Bitten and struck. And his car
Plumed a dust ball that rolled on the land
Until that too was gone.
All of it left to his wife save a note –
Took only his hat and jangled self.

     *   *   *

Was it an ache to know everything,
Even his own misery? Was it his waiting
For all things to end, even his dignity?
Or some kind of creation,
A dark out of brightness,
Despair out of hope,
Needed completion, obsession?

82

In the darkness of dreams, the night dancer
Returns to my room, smelling of night myrrh
And earth, glides from the curtains,
A shadow in shadows, uncoiling her flickering tongue.
"Lightness and darkness," she whispers.
"Each has its opposite. Life and the endless nonlife."
And she lies down beside me. I'm aghast
At how much she has changed. And she touches
My thighs with cool hands, kisses my neck
With cool mouth, gown shroud
Envelops me. Floating, I think
Of the smell of my groves,
Ebbing warm eddies and star gazer lilies,
The sound of the wind sweeping the beach –
All I will miss in her unending night.
I can feel myself shrinking. Is it over so soon?
I touch her cool belly,
Her breasts like ice sculptures. I think
Of the heat of the summer. She shines her cold moon
On my body, caressing my creviced dry skin.

She is the ending.
I open my mouth.
I will descend.
Is this some new kind of love?

83

The time of my singing grows shorter and shorter.
I know this by shadows and flesh,
By the size of Abbas's grandchildren –
Dreams of the future are now in the past,
And I dream of myself as a much younger man.

How should I breathe, in these last breaths of air?

When I see light on the sea, should I say:
This is the light that will always be?
When I see poppies that shiver in wind,
Should I say: This is the wind that will always be,
This is the flower born over again?

Should I imagine that ends are beginnings,
That all of it is as it was the first hour?
Or should I put time on the scale, like Lavoisier:
This much the weight of the ending,
And this much the weight of a life.
This much is wasted,
And this much will count –
So many years lost in these hot afternoons.

  * * *

84

It will pass quickly, so quickly, yet here I am
Stirring my tea. Abbas knocks at the door,
Calling me. What can I say in this dwindling time?

84

"Were you calling?"

"No," says Abbas,
"And you don't look so good."

I walk to the piano and play a malouf,
Song that my mother taught,
Qasida poetry, weaving in slow syncopation.
Abbas slowly closes his eyes, smiles,
Lets his cane slip and fall to the floor.
"Yes," he says, whispering,
"You still have faith."

On I play, threading the melodies,
Fragments of verse, off keyed piano.
Abbas leans and sways,
Smiling his old man's smile,
Yellow teeth.

They never wrote to me after I left,
Then or years later,
Not she, or the children –
Was it desertion, or saving my self?

85

It's Abbas who does not look so well,
Wheezing, unsteady, asleep more each day –
Goes to his room to lie down,
Leaving his sandals outside the door.

I take up my pen, turn on a lamp
In the watery light, dusk -
What is this feeling of otherness,
Memory smeared,
Years in diagonal stripes?
Odors roll in from the wet rolling sea,
Abbas's deep breaths
Keeping time with the clock in the hall,
Olive trees fade in the fading light, dreaming,
And I too fall into bed.

86

I dream of the fields of unplanted people
And wait for my seeding. A hollow wind
Skitters across the bare ground
While a pair of dark birds wings through the air,
Waiting to pluck up the seeds.

To my fellows unborn: We must wait, wait,
Through the flutter of wings,
While the soil is made rich with our waiting.
For we cannot choose

The precise time of our planting, or even
The century. Sleep, in the undreaming
Sleep of the unplanted, the wait without mind
Of the seed.

To all the unborn: Can we feel the millennia
Pass while we wait for our planting?
Can we sense the destructions of cities,
The smoldering of stars? And the hope?
Is there movement and shape
In the space of our eggs?

And when we are born – from the uncounted
Seeds that will sleep on forever –
Will we say: "This is the way it could be –
I might do this, and I might, and I might
Smooth out this crease at this moment,
Or close the blue gate to the garden."

Or will the uncountable futures be trapped
In the past, choices transfixed in the moment
Of planting, and even before, in the flesh
Of the unplanted seeds.

Was there some choice that I wasted,
Crazed, shattered with insults,
My children begging me, bones
And the breath of me?
Or was there no choice from the start,
Born to a singer and merchant,

To corroding wealth,
Butterfly pinned in a case?

87

Yasmine, Abbas's daughter, brings baskets
Of eggplants, bread, skewers of chicken and lamb.
Her eyes are Abbas's eyes, silky dark hair, tall.
"Should he come home?" she asks.
"This is his home."

Day after day he has stayed in his bed,
Too weak to rise, calling for people
From some other time. I pour tea
In his dry dune of a mouth,
Hold his limp hand, sit
While he sleeps through the day,
Shadows inch over the floor tiles,
Sea heaves of his breath.

88

A fine silt of sand slips over tables and floors –
Desert storm sifting its way through the house
Even with windows shut. Weeks pass,
Perhaps months, with the voice of the wind,
Shuttered light, food eaten from cans.

After he died, minutes after,
I touched him – his skin felt the same,
Scabrous, with heat rising up to my hand,
Even a pinkness. His hair smelled the same,
Odor of apple. I bent to the bed,
Beating my chest.

Waited for what, I don't know, waited
For time to stop, sound of infinity,
Glimpse of my own nothingness.
But all was the same – light from the lamp,
Shoes in their places, the photographs,
Sweets on his dresser, the whisperings,
Footsteps, the beads of the curtain,
And him just as he was –
Except that the life was gone.
Memories, stories all vanished
In seconds. A world in him then, and now nothing.
How did his atoms once talk to me?
Atoms feel pleasure, and thirst?
Minutes ago. And now they're just atoms,
Material shaped like a person, his face,
Grouped in a mass on the bed.

90

I bathe him as he bathed me years ago,
Rub him with dimpled cloth, lemon and soap,

Wrap him with three turns of a sheet,
Jasmine perfumed.
Abbas, dear Abbas, fare well my old friend –
But I am alone.

In the curved hallway, his daughters are draped in black,
Beating their chests silently.
Husbands sit sharing a pipe, also in black.
Villagers stand on the terrace
Reciting prayers, asking forgiveness.
Gently, I touch his gray stubble, his face –
I'll keep his worn sandals and cane.

Then to the grave, hole in the earth,
Head facing Mecca,
Three handfuls of soil, fragrance of orange trees –

Remember the reeds that you taught me to criss-cross,
Remember the letters you wrote me,
Remember the candle at night.
Why did you leave?

91

Moons rise and fall splitting the night –
Sleepless and weak from not eating,
I sit on my terrace and gaze at the sea,
Follow the curling dark waves,
Dark lung of water to dark lung of sky,
Light of a ship in the blackness,

I wait for your shore,
Sound of your breathing,
Your laughter, your prayers.

I talk to myself, answer myself.

Am I alone in this omnibus?
I must speak clearly as if you could hear
Words making sentences, narrative –
Traveler who lived
In this ocean of mystery,
World that destructs
In each second, then reforms itself,
Laws of mechanics and force, numbers
And world without numbers –
A theater of meaning, but what?

What is this crumbling dry villa,
These empty rooms, orange trees and groves,
Paper and pen, memories?
What are these sandals,
Still tinged with your smell?

Tell me one thing that is true.

I am bent between something and nothing.
Is this my hand, jumble of stubs
Pulsing with sap – hand, or the thought

Of a hand? Is this a grave of fresh soil?
Grove of my father?
Cracked steps to the sea, windows,
This villa, the rooms of my childhood,
The voice of my mother, footsteps of Abbas?
Might the land be a dream, vanished
In one wakeful moment, like breath,
Like a wrinkle of wind on the sea?

93

Where are you, Abbas?

Is there some essence of flesh that is fleshless,
Some bone that is boneless, a breath
Beneath breathing, a center that has no periphery?
Is there some substance without Newton's mass?
Is there a world of nonmatter
Within the material world?

Water and air could not spill from its
Emptiness. Sunlight could never reflect
From its surfaces, space without objects
And space without space.
Still, it might redeem –
Lightless and soundless yet
Full of a thing that cannot be named.
Is that where you are?

     *   *   *

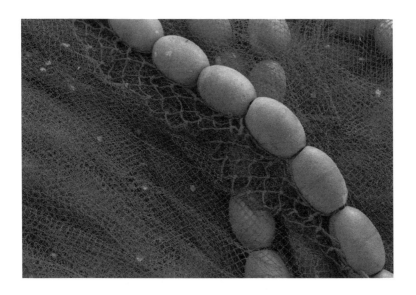

So fragile would be this invisible world,
Like a ghost without whiteness,
But strong, like the spinnings of spiders,
A gossamer kingdom without towers or walls,
Indestructible,
Large as a cosmos and small as a seed.

Is this the small whisper I hear in my night?
The invisible spurt of invisible blood?
How can I touch it?
I want it to know my small hands.

94

Here on the beach, as the fisherman patches his nets,
Pausing to gaze at the moon on the sea –

Here, as I walk through the garden at dawn –
Here, as the light again slides
Through the shutters, I wonder if it will all happen again:
The primordial explosion, the atoms
Vibrating in decimals, galaxies
Flung into blackness,
Each gesture, each word that is spoken,
My mother's sweet singing,
The small town and prefect,
The oud player's pain –
Even now, at this instant, I touch this spined puya –
Will all of it cycle through ends and beginnings,
Repeating, repeating each fragment of time?
I strain to remember the previous world.

Or is it just once,
One small experiment,
Universe shot from a cannon?
Exhausted, the stars will be ashes,
The coastline will slide to the sea,
And this breath will not happen again.
Feel it, it burns and expires.

Who will record what we do and say?
Who will remember these pale dunes at dusk?
Pages will crumble like cities,
The pictures and bits will decay,
No one will know.
I must hold these scarce words in the air.

95

Should I not go to the boats on their anchors
And listen to waves rolling by, carefully
Listen for something,
Necessity, aim in the aimlessness?
Where is this moment against
The flashed arching of time?

I say to myself: I am not ready to hear
What cannot be heard,
But the sound may be soundless.
And I am not ready for faith,
But the vision may come to the faithless.
I am not ready . . . but
I will do these few things
And sing out with the cosmic sensorium.

96

I am wakened from sleep by a voice in the night,
Distantly calling – the oud player's wife
Calling him, cursing him, calling him.
Says she's been rambling for months,
Town to town – "Where is that rotten man?"

She looks behind trees, shambles and moans
Like a woman in love.
All I can see is her dark flowing veil,
Night silhouette.

    *   *   *

I hear music,
Not oud but piano,
The yearnings, then torrents,
Melodic and sinuous,
Castles of things without size,
Music that lifts me and hallows,
And I am in tears.
And I need to remember
The journey of footfalls and winters,
The soft settling of light in the late afternoon –
Crushed and becalmed
In this music,
Alive, I'm alive,
And I still want this world,
Lurching and sweet.

Where is the oud player's wife?
I search in the dark.
I must tell her. I call to her:
Listen, your anguish and hope,
It is life.
But she's gone.

97

Travellers who travel in this slice of time,
Come with me now on this last voyage of flesh,
All of us seeking a closing, completion,
A chime of the bell,

End from beginning,
An end that lives endlessly even in ending.

But there's no completion in dawns and their blood flush,
Or beatings of hearts charting their hours,
Or chemists who measure their structures and grams,
Or the icebergs that groan
As they split in the sea,
Or the fields of the night-blooming jasmines and lilies,
The blue flowered flax that is pulled with its roots –

There's no completion in herons that wade in the water,
Or pheasants that flit in the brush,
Or the drooping persimmon, the black coastal mangrove –

There's no completion in engines that whirl in the night,
Or the steel girded towers that puncture the sky,
Or the roadbuilder's concrete and scab slabs,
The clicking and whining of office machines,
Even the silence surrounded by noise –

There's no completion in marriage,
Or unending end of a marriage,
Or waiting for what never comes,
Or this villa that dies by the sea.

There's no completion in shadows that fall on the terrace,
Or grey winter mornings,
Or beige summer light,
Or the phosphorous glow of the watery algae.

Or the book that's read over and over again,
Or the word that is spoken,
Or painting with hog bristle brush,
Or playing my mother's piano,
Or writings of Lao-Tsu, Omar Khayyam,
Or the close of the door to the terrace, just so,
Or the close of the shutters at dusk –

Or the infinite digits of *pi*,
Or equations for gravity,
Perfect ellipse –

Or the bride who unbuttons her rumpled white dress,
Or the family that breaks
When the father returns to his country,
Or boys who collect the tin cans on the street,
Or the bank tellers sorting the coins and the bills,
Or the soldiers who crawl through the sludge
With cocked guns,
Or the writers of letters that wait in locked drawers,
Or the lawyers and pilots and teachers and dragglers –

There's no completion in patterns,
For patterns are constantly restitched in new patterns.
There's no completion in history, which kneels
Bare and mute at the feet of the future.
There's no completion in mind
With its unending halls,
Or electronic minds that have no beliefs.
        *   *   *

There's no completion in seasons, repeating repeating,
Or Earth as its spin traces loops through the stars,
Or the Sun as it slowly consumes itself, fire on fire,
Or Space as it twists and expands in the dark –

Or the pitiless ticking of clocks,
Or the withering of snapdragons after their seeds,
Or my crippled dear Abbas bent over his cane,
Or the hand as it cleanses the wound,
Or the kiss that brings life to life,
And then later to nothingness.
There's no completion in nothingness.

99

Where can I sit on this trifle of dirt
That revolves without aim in the blackness
Of space? If I ask, will the asking,
Unanswered, come round on itself –
Even in emptiness make a thing whole?

My companions are gone – Newton and Darwin,
Al-Haytham and Einstein, Lavoisier. Lao-Tsu,
Rembrandt and Omar Khayyam.

My mother and uncle, my lost wife and children,
Abbas and his daughters and sons –
Whom can I love who will not pass to nothing,
When all pass to nothing, along with this song?

    *   *   *

Is there nothing and nothing,
A drivel from dry sea beds
As time slides to an end?
It cannot be so.
I must reach out to what I'm unable to grasp,
Reach out to what I want to believe,
And my mutterings slip from my lips,
Faintly, and faintly dissolve in the air,
And my small room is soundless again.

List of Photographs

Printed and bound by CPI Group (UK) Ltd, Croydon, CR0 4YY

21/10/2024

01777083-0004